Lean
vs.

Agile

vs.

DESIGN
THINKING

Lean
vs.

Agile

vs.

DESIGN
THINKING

What you *really* need to know
to build *high-performing* digital product teams

Jeff Gothelf

Issued in print and electronic formats.
ISBN 13:978-1541140035 (paperback).
ISBN 10:1541140036 (epub).

Editor: Jeff Gothelf
Cover design: Michel Vrana
Interior design and typesetting: Jennifer Blais
Illustrations: Remie Geoffroi
Author photograph: Dailon Weiss

Published by Gothelf Corp
728 Prospect St
Glen Rock, NJ 07452 USA
+1.551.579.2312
Printed and bound in the United States.
1 2 3 4   20 19 18 17

"ONE DOG'S GOING ONE WAY. ONE DOG'S GOING THE OTHER WAY. AND THIS GUY'S SAYING, 'HEY, WHAT DO YOU WANT FROM ME?'"

"OUR TECH TEAMS
ARE LEARNING AGILE.
OUR PRODUCT TEAMS
ARE LEARNING LEAN,
AND OUR DESIGN TEAMS
ARE LEARNING
DESIGN THINKING.
*WHICH
ONE IS RIGHT?*"

There's an unforgettable scene in my favorite movie, Goodfellas, where Joe Pesci, Robert DeNiro and Ray Liotta, after an *eventful* evening, pay a late-night visit to Pesci's mom. The purpose of the visit was simply to pick up a knife to help them get the "hoof" of a deer they struck off the bumper of their car (or so they say). Despite their best efforts to get the knife and leave quickly, she convinces them to stay for a bit and have something to eat.

During the late-night meal, conversation turns to a unique painting they spot in her kitchen. It's of a bearded, white-haired man in a boat with two dogs.

They pass the painting around remarking on its uniqueness and the seemingly odd scene it depicts. When the painting gets to Pesci, he utters one of the most memorable lines of the movie, "One dog's going one way. One dog's going the other way. And this guy's saying, 'Hey, what do you want from me?'"

The men explode into laughter remarking how the guy in the picture reminded them of somebody they ran into earlier that evening (and who currently occupied the trunk of Liotta's car).

I experienced a similar, albeit less "Goodfellas," moment with a client of mine in 2016. We were preparing for an upcoming training workshop focused on coaching a cross-functional team of designers, software engineers, product managers, and business stakeholders on integrating product discovery practices into their delivery cadences. During our conversation, they said to me, "Our tech teams are learning Agile. Our product teams are learning Lean, and our design teams are learning Design Thinking. Which one is right?"

The client found the different disciplines at odds because these seemingly complementary practices forced each discipline into different cadences, with different practices targeting different measures of success.

The engineering teams were focused on shipping bug-free code in regular release cycles (many teams call these "sprints," though the term "release train" has grown in use with the rising popularity of SAFe—the Scaled Agile Framework). Product managers were most interested in driving efficiency, quality, and reduction of waste through tactical backlog prioritization and grooming techniques. The rationale behind these practices emerged from Lean thinking but, in practice, had nothing to do it. Not to be left out, the designers sought to bring the customer front and center by validating problem-solution fit with Design Thinking activities, yet their activities were perceived as lengthy, up-front research and design exercises that delayed product launch. Each discipline was working through its own ceremonies and tactics, targeting an ideal state of success unique to them. The collaboration, shared understanding, and increased productivity they were all promised was nowhere to be found.

"So what?" they asked. Each discipline should work in whatever way is best for them, right? No. Without a clear understanding of the problem, product managers optimized backlogs of work based on gut instinct and subjective preference from stakeholders. Without a clear understanding of the customer, engineers focused on simply shipping features—the more, the better—without a sense of whether they helped to solve a real customer need in an effective way. And without any sense of the feasibility or strategic alignment of their prescribed solutions, designers came up with ideas that never stood a chance of seeing the light of day.

HOW DID THIS COME TO BE?

U nfortunately, this was not the first time I'd come across this situation. Agile, conceived originally to bring a more responsive, evidence-based, customer-centric way of working to software engineering, has been productized in recent years. This productization has changed the intended goal of the Agile work philosophy. Instead of focusing on the responsiveness of the software engineering teams,—i.e., how quickly a team could react in the face of uncertain market conditions, software complexity, evolving customer behavior, et al.,— Agile is being "hired" today to drive velocity. The only goal that matters to most Agile software teams is the efficient delivery of high quality code. Coaches, trainers, workshops, books, webinars, and blogs extol the efficiency virtues of Agile, promising greater predictability, throughput, and productivity by limiting customer feedback to the stakeholder, not the end user, while relegating design to nothing more than a final coat of aesthetics.

This is, of course, a ludicrous concept given the unpredictable and complex nature of software. Adding in "distractions," such as design, experimentation, or continuous learning, only increases the amount of time it takes to get code to market. The seemingly endless parade of Agile coaches and trainers have mostly chosen to focus narrowly on these ideas of velocity and efficiency, ignoring the other elements needed for successful digital products, and then marketing their services to the tech managers they believed were most likely to buy. Then, well-meaning managers trained their teams within their discipline and never thought to look beyond, because their new coaches never suggested it.

Software engineering is not alone in this. Many organizations are growing and ramping up the skills of their product management teams. To this group, Lean Startup in the Enterprise is the process of choice.

Favoring a regular cadence of experimentation, product discovery, and learning, Lean Startup has found mass appeal with large organizations seeking to recapture the energy and agility of smaller, disruptive companies. However, dig into what most organizations know of Lean Startup, and you'll likely end up where most Lean Enterprises do—the MVP. Standing for Minimum Viable Product, MVP has become one of the most powerful and bastardized phrases in modern product development. As product managers strive to steer their engineering colleagues to build only what they need, the conversation inevitably nets out at what most would consider Phase I of the product, with the reality of a future Phase II never a sure thing. Despite Lean Startup's core foundation on Agile principles and rhythms, few product teams work to integrate these two practices.

Finally, we come to designers. The last 20 years of digital and physical product innovations, led by companies like Apple, Netflix, Tesla, and Nest, have vaulted design into the mainstream and the boardroom. Designers were thrilled. Years of being relegated to "making it pretty" were finally waning in favor of greater strategic influence. The legendary design firm IDEO took advantage of the rise in popularity of design to spread their way of working to the corporate masses. Design thinking took flight. Using the toolkit designers had honed for decades, but applying it to broader strategic business problems,

Design Thinking was heralded as the ultimate advocate for the customer in product organizations of all sizes. Executives attended seminars. Teams attended workshops. Certificates were printed. Sketches were made. Post-its were consumed. And at the end of it all were amazing concepts that heralded future states of products and services that failed to align strategically to corporate goals and were often too complex to implement. The work, executed by the book, failed to include a broader spectrum of collaborators. Existing technical limitations, risk-averse cultures, and cultures more motivated to protect their bonuses than improve the customer experience were never considered in the conception of these designs. They never stood a chance.

With each discipline now displaying certificates on their cubicle walls, it was time to get to work— together. The net result? Confusion at best. Chaos at worst. Teams that were supposed to start building trust through cross-functional collaboration were now at odds about how to start. Should we run a design studio? Should we build an MVP? When do we ship code? Who leads? How do we measure success?

Tech teams focused on increasing velocity. Product teams focused on reducing waste. Design teams wanted lengthy, up-front research and design phases to help discover what the teams should work on. Very quickly they found themselves pulling away from each other, as opposed to collaborating more effectively.

And their managers, thinking they were doing the right thing for their staff, were left like the Goodfellas painting—one team going one way, one team going another way, and the manager in the middle saying, "Hey, what do you want from me? (I trained them in modern methods.)"

Let's take a (slightly) deeper look into each one of these methods to understand its genesis, core philosophy, and common ways each manifests in most companies.

AGILE

"WE VALUE
RESPONDING
TO CHANGE
OVER FOLLOWING
A PLAN."

A gile was born of frustration. In the mid to late 1990s, software development was facing a crisis. Project after project, if they ever launched, failed to meet customer expectations, budgets, and prescribed timeframes. Despite the elegant Gantt charts that exemplified the waterfall planning style popular at the time, customers got software systems that didn't work well, were difficult to use, and didn't provide the value they were designed to. A group of forward thinking engineers assumed there had to be a better way. Seventeen of them spent a long weekend atop a mountain in Utah debating the best way to approach software development to prevent the failures they were seeing in practice.

At the core of the debate was uncertainty. The products and services being developed were not made of physical material. Their end state could not be predicted up front. The level of complexity was also difficult to nail in advance. Perhaps most challenging, the way customers engaged with these new systems was not guaranteed. They weren't manufacturing pens or automobiles. In those cases, there are clear specifications, costs associated with materials, clear use cases for the final product, and a limited number of ways those products would be consumed by customers. Software didn't provide that level of predictability; hence, using manufacturing management practices didn't make sense.

Instead, the authors of the Agile Manifesto approached software development with a philosophy rather than specific tactics. At the core of that philosophy is this phrase, "We value responding to change over following a plan."

Many things may change over the course of a software development project. Market conditions may shift. A competitor may launch a similar product before you've finished yours. The economy may face unforeseen stress that changes consumer buying habits or needs. The complexity of the service your team is building may prove to be far greater (or smaller) than anticipated. To deal with this uncertainty, the authors advocated working in short cycles and then pausing at the end of each cycle to reflect on accomplishments, learnings, new insights, and next steps. Does the accumulated total of this data indicate we should continue in the same direction? If so, let's work another short cycle in the same direction. However, if the feedback we've collected over the last cycle (or "sprint") indicates our current

trajectory is flawed, we are obligated to change course—
that is, to *be* agile.

In the years since the Agile Manifesto was written,
the uncertainty associated with software development
has gone up, as the costs of creating it have gone down.
Perhaps nothing has had more of an impact on how
code is written and delivered in the last five years than
the DevOps movement. Without getting too technical
(I'm not, after all, a software developer), DevOps allows
companies to ship code to customers in a continuous
state. Launching a new piece of code becomes a non-
event. It is something that happens every day and, in a
growing number of companies, many times per day. As
easily as an organization can launch new code, it can roll
it back, tweak it, and launch it again.

Why is this revolutionary? Because it changes the
pace with which teams get feedback on the efficacy of
their product. As soon as new code is shipped, customers
can start using the new functionality. That usage yields
feedback that allows teams to determine how well that
functionality met customer and business needs and if it
needs to be improved.

With delivery of code becoming a non-event, the
nature of software has changed from a static piece of
code delivered via physical medium (e.g., CD) to a
continuous, dynamic system that is regularly optimized,
augmented, and improved. In many ways, DevOps has

enabled even greater agility in software development teams since they can now respond to change as quickly as they uncover it.

Ask anyone in a tech role in a large company what development methodology they use, and in almost all cases they'll respond, "We're doing Agile." The challenge with Agile is that the authors never addressed how to implement it at scale. The Agile Manifesto speaks to team-level practices. Incorporating continuous feedback at that level is relatively simple. The risks are known, as are the consequences. Scale that out to 10, 100, or 500 teams and the potential for organizational chaos grows exponentially. Teams start optimizing locally, not concerned with their colleagues focused on a tangential success metric. Sometimes these simultaneous initiatives are complementary. Other times, optimizing flow in one spot detrimentally affects flow in another. Efforts are duplicated across teams that don't communicate with each other, leaving the expected efficiency that scaled Agile promised in the dust.

Never fear, consultants have solved this. Enter the Scaled Agile Framework or SAFe, for short. Driven by a set of complicated diagrams, railroad metaphors, and a hoard of certifications, coaches, and trainings, SAFe promises large organizations the responsiveness the Agile Manifesto authors promised. Except, as with most things Agile, it treats the other disciplines—

product management, design, marketing, et al.—as support staff for the engineers. How then, do teams reconcile the drive for velocity, Agile at scale, DevOps, and SAFe, while ensuring meaningful contributions from non-tech disciplines?

LEAN
STARTUP

THE QUESTION EVERY PROJECT SEEKS TO ANSWER IS NOT, "CAN WE BUILD IT?" IT'S, *"SHOULD WE BUILD IT?"*

Founded first as a thesis after several failed startups, then a book, and now a global community and movement, Lean Startup has swept through corporations looking to recapture the momentum they had as smaller, more nimble companies. Spearheaded by entrepreneur and professor Steve Blank, and perhaps even more notably by author and entrepreneur Eric Ries, Lean Startup posits that every project you take on is an experiment. The question that experiment seeks to answer is not, "Can we build it?" it's, "Should we build it?" And, if the answer is yes, "Can we build a sustainable business model around it?"

Ries advocates for a cadence of experimentation and iteration that only warrants further investment if each experiment yields evidence that more work should be undertaken on this initiative. Short cycles (sound familiar?) are used to run these experiments and, once data has been collected, teams reflect on whether to persevere with the project as planned, pivot in a new direction, or kill the idea altogether. The determining factors for this decision are customer behaviors. Did we get an indication that our customers want this feature? Did they indicate they would pay for it? How do we know? What does that tell us?

These are just a few of the questions Lean Startup methods help answer. And yet, again, ask those same anyones at large corporations if they've heard of Lean Startup, and they'll say, "Yep, we're building MVPs."

Sigh.

The MVP was designed as a tool of Lean Startup to help teams answer two questions:

1. What's the most important thing we need to learn first on our project?
2. What's the least amount of work we need to do to learn that?

The first question speaks to risks. For the team's current initiative, what's the thing most likely to make it fail? Once the team reaches agreement here, they move on to the second question which addresses the real meaning of an MVP. It's not an attempt to be lazy, but it *is* an attempt to do less work before we commit to a full build-out of the feature. If we learn from our MVP (or experiment) that our intended solution does not actually solve the problem successfully, why would we continue to work on it?

Unfortunately, with the push for velocity on most teams, the term MVP has lost its original intention and

meaning and is now largely a stand-in for Phase I—
or, what is the fewest number of features we can get away
with and still ship this product?

Experimentation is not something for which teams
are typically rewarded. So, while many teams are
shipping less software, at least initially, they are not
leveraging that initial release to learn and optimize.
Instead, they are moving on to the next set of features in
their backlog and the next "MVP."

Lean Startup, although predicated on Agile's short
cycles, still faces an uphill climb integrating into most
companies' cultures of delivery. The main challenge
is that, once again, it introduces uncertainty into the
process. If we learn that something we're working on is
no longer the right thing to do, but we've committed to
a date and a roadmap, what do we do now? And, if we
can't make these kinds of adjustments at the same pace
we can learn them, are we truly Agile?

Lean Startup has been largely the domain of product
managers in large companies. It is typically their job to
figure out what the teams work on and in what order.
They're responsible for figuring out the scope of each
release and for shepherding the team to those outputs.
Since course correction is often frowned upon, any
Lean Startup-style experimentation is often done up
front, once, and then taken as "proof" that the team
should execute the plan as written. It is a rare corporate

software team that actually has the autonomy to kill a feature and is willing to do so.

To combat the pushback on Lean Startup in the Enterprise, some companies have launched innovation labs. Inspired by Clay Christensen's *The Innovator's Dilemma*, these are stand-alone business units immune (theoretically) to the day-to-day obligations of a traditional business unit and tasked with charting the future of the company's product offering. They don't have a P&L. They're not expected to return a profit. They're meant to be the "creative ones."

Their modern offices, bereft of cube farms, feature chromed-out espresso machines, Italian foosball tables, open work spaces, and lots and lots of bean bag chairs. What they don't often get is a strategic mandate and a clear path for transitioning a product idea, once validated, into an integrated production track with the broader organization. This isolation from the rest of the organization, meant to provide creative freedom, is why most innovation labs rarely yield anything of value. Separating out the critical thinking from the production capability invariably creates conflict during the handoff. The production teams don't have the same enthusiasm for the freshly hatched idea the innovation lab is handing over. Their backlog is already packed, floor to ceiling, with work they're on the hook to deliver, relegating the new initiative to a launch date so far in the future it may not even make sense to ship it at that point.

How then do we bring the immediacy of validated learning to inform our Agile delivery cycles and help product managers become more than glorified project managers?

DESIGN
THINKING

"ARE WE SOLVING
A REAL PROBLEM,
FOR A REAL CUSTOMER,
IN A MEANINGFUL WAY?"

Agile helps us deliver work in regular cadences. Lean Startup helps us determine what to focus on. How, then, do we determine whether we're actually working on something of value? Enter Design Thinking.

Popularized in the 1990s by legendary design agency IDEO (though around for decades before) and almost impossible to nail down in one definition, Design Thinking teaches teams to take an empathetic look at the customers they are building products for to understand the core needs being addressed. Then, through a series of facilitated brainstorming exercises, to come up with a set of solutions that not only meets those needs but is technically feasible and viable for the business. At least, that's how it's supposed to work.

Despite the executive training courses at Stanford's d.School, 2.2 million presentations on Slideshare.net and, at the time of this writing, over 2600 results on Amazon.com for books on the topic, customer empathy is hard to find in large companies. Designers, brought in to infuse some "Apple"-ness into the organization, are often left with the task of user advocacy. Buoyed by the frequency the phrase "design thinking" is used in these companies, these designers create workshops, brainstorming sessions, and design sprints and other activities designed (see what I did there?) to increase the team's empathy for the customer. The output of these sessions manifests as sketches, photos of Post-it note clusters, and whiteboard scribbles, discernible, if at all, only to the person who made them.

At best, these sessions help teams build a shared language and common sense of purpose for the upcoming project. At worst, they're perceived as a waste of time that could have been used to write code. Design Thinking practices are there for teams to use continuously to ensure projects stay on track and are adjusting properly to the learning derived from our Agile practices. However, most teams, having failed to arrive at an epiphany during their first brainstorming session, rarely employ these practices throughout the lifecycle of the project. The continuous nature of software, which should yield a cadence of continuous learning, is once again retrained on shipping bits to customers without the benefit of team-wide shared understanding of the core customer need.

SO,
WHICH PROCESS
IS RIGHT?

YOUR JOB IS TO
PICK AND CHOOSE
THE SPECIFIC ELEMENTS
FROM EACH PRACTICE
THAT WORK WELL
FOR YOUR TEAMS
AND THE BRAND VALUES
YOU'RE TRYING
TO CONVEY.

With such differing practices, motivations, and measures of success, is it any wonder organizations find it difficult to reconcile these processes and create highly-productive, creative, balanced teams?

There are valuable components of each of the various processes teams are trying out these days. As an organization seeking to leverage the benefits of continuous improvement and a software-based service offering, your job is to pick and choose the specific elements from each practice that work well for your teams and the brand values you're trying to convey. In my practice, I've found that a few core practices serve as effective starting points. I recommend:

1. **Work in short cycles** —Software is complex and unpredictable. So are people. Assuming you know exactly how people will respond to change is the same as assuming you can predict the end state of a piece of software— impossible. Agile transformations are often taken as wholesale, top-to-bottom initiatives.

Since you cannot predict how people will react to this change, this approach is highly risky. To reduce the risk of implementing sweeping process changes that fail, take small steps. Pick a new idea or practice and try it with a subset of your teams. This new practice shouldn't be handed down as a permanent change. Instead, present it as a "process experiment." Let the team try it for a limited amount of time (e.g., two sprints) and see how it works. If it fails, the team has invested very little time and effort in this change, but (hopefully) they've learned a lot. If it succeeds, the team keeps the practice, improves on it, and the organization rolls it out to more teams.

For example, here's a small experiment to run with your team to increase their amount of customer exposure (a value lauded by Agile, Lean, and Design Thinking): Every two sprints, the entire team goes out, together, to observe customers using your product for a half day. The specifics of the outing—whether simple observation, planned usability test, or customer interviews—are largely irrelevant. The goal is to get members of the team who don't normally come into contact with customers to do so on a regular basis.

2. **Hold regular retrospectives**—Retrospectives are the heart of continuous improvement. They are a regular opportunity for the team to consider their current practice, evaluate its efficacy, and determine how to progress. Many teams who claim to practice Agile don't hold regular retrospectives. Initial retros are uncomfortable and often feel like nothing more than venting sessions. Even teams who hold these sessions regularly often generate so much improvement material that it feels like very little is actually getting better.

    At the end of each sprint, encourage your teams to get together for an hour, review

what worked well during that cycle, what
didn't go well, and commit to improving
one or two key things each time. Oftentimes
this will feel awkward at first but, after a few
retrospectives, teams begin to open up more
freely and address the core issues hurting
their collaboration. If this fails to occur, offer
teams an outside facilitator for these meetings.
Someone without anything at stake in the
retrospective will do a better job probing for
root causes and solutions.

3. **Put the customer at the center of everything—**
The customer is the person who has to use the
product. It's really as simple as that. Without
customers, there is no business. There is
no team. There is no product. Agile, Lean,
and Design Thinking prominently tout the
supremacy of the customer in all activities.
Yet, each methodology offers a different
definition of who the actual customer is.
While Lean and Design Thinking generally
align on "the person buying and using the
product" as the customer, Agile often refers
to product owners or "the business" as the
customers. If they're not using the product,
they're not the customer.

If you're struggling to get alignment as a
team, focus on delivering customer value. If
there is debate within the team about what to
work on, how much more time to spend on
something, or whether the product is headed
in the right direction, consider asking the
following questions on a regular basis:

*How do we know we're shipping something
users care about?*
This question will quickly reveal whether or
not your teams have a good feedback loop
with their customers. If you don't know

whether your customers value what you're shipping, you've identified a critical gap in your understanding of your work.

How do we find out?
This question addresses the customer feedback gap identified in the previous question. Do you need to go out into the field to meet customers? Do you need to implement an analytics system? Is there a conference you can go to where your customers regularly congregate?

How does that affect what we prioritize?
Once the feedback loop is in place, the team will need a method for incorporating it into their decision-making process in a timely fashion. Holding regular customer development

debriefing sessions, as well as bringing these findings into your iteration planning meetings, is a good place to start.

4. **Go and see**—Go to the *gemba*. This is a Japanese phrase born of the Lean manufacturing movement that literally means, "Go and see." The point of the phrase was to get managers out of their offices and onto the manufacturing floors to see how the teams were working. The same holds true for software teams. As a manager, it's your responsibility to regularly walk around, talk to your teams, and ask them what's working and where they're struggling. Bring those learnings back to your management meetings and share with your colleagues. Patterns that yield good outcomes, regardless if they're Lean, Agile, Design Thinking, or some other methodology, should be amplified and scaled. Those that are causing problems should be diagnosed and remedied or discontinued. Working from recipes is a good place to start, but once those systems are deployed within your unique organizational context, they will inevitably morph to the needs of your specific environment. Don't shun successful practices

simply because they don't fit neatly into one of these pre-defined recipes.

5. **Balance product discovery with delivery work by only testing high-risk hypotheses—** It's a rare occasion when a team is tasked solely with learning. Inevitably, your teams have to ship product to market. Often, product discovery work is perceived as a block to shipping code. "Oh great, another brainstorming session." "How long do we have to wait until the research is done?" "Why do we have to test everything?" This perception is exacerbated when each discipline on the team wants to practice "their flavor" of product discovery. All of a sudden, you have the product managers running one set of activities, the designers advocating for a second set, and

the engineers itching to get out of all of them, in the hopes of one day, eventually, writing code.

The reality is, teams can't test every task in their backlog. In fact, they shouldn't. There are assumptions in every backlog that are riskier than others. Prioritize these assumptions based on their perceived risk and their perceived value. Everyone on the team should weigh in on risk, because each backlog item may present a different type of risk. For example, an item may have technical risk associated with it. An engineer would need to provide that perspective. Alternatively, that same item may involve design complexity that is deemed high risk. A designer will give that point of view. Risk could also manifest as market risk—an opportunity to provide services to a tangential audience from your core. Product managers are particularly good at providing competitive and market analysis in this situation.

The items that qualify as high risk and high value are the only ones that should be subjected to product discovery efforts.

Once those risky assumptions are identified, determine what the most important thing you need to learn about them is, and

then leverage the most appropriate discovery technique to learn that. For example, if technical risk is most important to verify, perhaps a tech "spike" (an Agile technique that timeboxes a technical research effort) is the right approach. If you need to understand whether an idea has value before you build it, you can use an A/B test or a landing page test (a classic Lean Startup tool) to learn that. And if you're unsure whether the problem you're solving actually exists for your customers, a contextual site visit (a Design Thinking tactic) might be in order.

By limiting product discovery work to the riskiest items, you ensure the team is delivering features while it's learning. At the same time, you're only using the tool you need to maximize your learning in the time that you have.

6. **Do less research, more often**—User research has been around a long time. As a tool in the arsenal of design teams, it was traditionally wielded with the subtlety of a giant hammer. Regardless of what was being tested (or what the team needed to learn), it often required at least two days offsite, in a facility stocked with candy, a dozen test participants or more, and a broad array of team members checking in and out as their calendars dictated. With a competent moderator and reasonable testing script, almost every major issue was revealed within the first five test participants. Every subsequent one yielded little additional value and cemented the perception in almost every attendee's mind that this was a colossal waste of time. The worst part? They were right. It *was* a waste of time.

   Now, don't get me wrong. I am a huge advocate for user research. However, having been the victim of many of these sessions,

I know how much waste is involved in each. In addition, I've witnessed firsthand how attendees, who were convinced that this was an essential use of their time, lost faith in the process and the technique.

To avoid this, I recommend continuing to practice user research with a cross-functional team in attendance, but simply do less of it, more often. Instead of testing 12 people, test three. Take the learning from that, and then do the test again the next week. Don't go offsite. Work in the office to ensure maximum participation. Most important, broadcast your findings broadly immediately after the test. Show the value of the exercise, reduce the commitment for participation, as well as the cost, and you'll find an increased level of organizational buy-in for this classic product discovery technique.

7. **Work (and train) as one balanced team—** The client quote that precipitated the writing of this book highlighted the discipline-based divisions at the root of that team's dysfunction (and that of others). Part of the reason each discipline is training in a different methodology is because, while their vocabulary may have shifted, the way they work hasn't. They continue to work in silos, handing off items from discipline to discipline. They are not working together on the same things at the same time. This is waterfall development with Agile language (some call it "AgileFall") and we've been actively moving away from it for years.

To combat this tendency, reconsider how you staff projects. The atomic unit of planning for any project is the team. The team is made up of designers, software engineers, and product managers at the very least. There is no "engineering team" or "design team" at the project level. These balanced teams provide the expertise, perspectives, and skillsets necessary for

all aspects of a project. They should be independent, autonomous, and empowered to make decisions based on the evidence their product discovery work uncovers.

When structured this way, it doesn't make sense to train different members of the team in different ways. There is no difference in the cadence of engineering, design, nor

product management on balanced teams.
Their efforts have to be coordinated,
aligned, and targeted at the same learning
and delivery goals. Balanced teams choose
the best parts of Lean, Agile, and Design
Thinking and apply them as needed in a
tight collaboration.

8. **Radical transparency**—Some members
of your team will have experience with
Agile. Others will be experts in Lean
Startup, and yet others may have gone to
a Design Thinking course. When
approached with a new way of working,
many people struggle. They've learned
how to do their job a certain way. They're
good at it, and it's served them well in their
career. Why should they change? How
does this new way of working make them
more successful at their job? This is where
transparency comes in. As new processes
are rolled out, ensure that it's clear why
you're trying something new. Let people see
how the process plays out with other teams.
Hold open retrospectives so that others
can get a sense of where a new process, like
Agile or Lean Startup, struggles in your
company. Be clear about what success looks

like, make sure everyone is aligned to that criteria, and identify how you'll measure it. If certain aspects of a process don't fit with your organization, be clear about why you're abandoning them and what you'll do instead.

Post success metrics in public places around the office, and show how progress is (or isn't) being made against those metrics. This can be done with real-time monitors showing data or, as one of my teams once did, you can print out the state of a metric every day you walk in to the office. Indicate how it changed from yesterday, and post copies of that all over the office.

In all cases, default to open.

9. **Review your incentive structure (and performance management criteria)**—This is perhaps the most important criteria for ensuring your teams choose the most productive amalgamation of these philosophies to work with. Teams will optimize the work they're incentivized to achieve. If you incentivize velocity, teams will work on getting more features out to market. If you incentivize learning, teams will build better product discovery processes.

These same incentives must be reflected in your company's performance management criteria. If you want to build collaboration and learning, employees must be assessed on the efficacy of their collaboration and their ability to build continuous learning into their work. For example, velocity is only rewarded if user satisfaction reflects value in the features shipped. Incentives like these force the kind of self-organization Agile teams are famous for. Knowing that their company values these behaviors motivates

teams to figure out which pieces of Agile, Lean, or Design Thinking will best help them achieve that.

10. **Make product discovery work a first-class citizen of your backlog**—One symptom that seems to manifest consistently across organizations is that the work that gets visualized is the work that gets done. Agile, particularly, has very clear directions, ceremonies, and rituals around visualizing work. This is why, in most companies practicing some form of Agile, you'll see physical boards or monitors displaying the teams' work in JIRA or other digital project management tools.

Agile advocates continuous learning. So does Lean Startup. Design thinking focuses on learning as well. Yet, there is no clear process or ritual linked to visualizing learning work. The delivery aspects of Agile get visualized, measured, and executed. In this situation, Agile "wins" because the (valuable) activities of Lean Startup and Design Thinking rarely find their way into the same tools as the Agile activities. The result of this is that this work doesn't get treated the same way as delivery work. It marginalizes the effort and allows it to be cut in the event of a time or scope crunch. Team members, often asked to meticulously track their time and effort on delivery tasks, are left to find time for discovery work "in the cracks" of their calendar.

To avoid this, product discovery work must become a first-class citizen of the backlog. It must be visualized along with the delivery tasks. It must be prioritized against them and assigned to specific members of the team. It must be tracked like delivery tasks, and the implications of the discovery work have to be taken seriously. In many cases, learning will reveal

gaps in your backlog or a poor prioritization decision. Changing your plans in reaction to these learnings is agility. It's the whole reason to adopt this way of working and is the key to building responsive teams and organizations.

HERE'S THE BOTTOM LINE

At the end of the day, your customers don't care whether you practice Agile, Lean, or Design Thinking. They care about great products and services that solve meaningful problems for them in effective ways. The more you can focus your teams on satisfying customer needs, collaborating to create compelling experiences, and incentivizing them to continuously improve, it won't matter which methodology they employ. Their process will simply be better.

AT THE END
OF THE DAY,
YOUR CUSTOMERS
DON'T CARE WHETHER
YOU PRACTICE AGILE,
LEAN, OR DESIGN THINKING.
THEY CARE ABOUT GREAT
PRODUCTS AND SERVICES
THAT SOLVE MEANINGFUL
PROBLEMS FOR THEM
IN EFFECTIVE WAYS.

ACKNOWLEDGMENTS

This book was made possible due to the amazing response of my online readers. It's through your continuous support, questions, push backs and discussions that we all get smarter and our community benefits. Thank you for pushing me to continuously improve.

In addition, I had a lot of help navigating the land of self-publishing for the first time from my friend and colleague Christina Wodtke who gifted me not only with her knowledge but with her contact list of excellent editors, designers and reviewers. Those folks include Cathy Yardley (developmental editor), Jennifer Blais (typesetter/designer), Michel Vrana (cover art), Remie Geoffroi (illustrations), and Shannon McKelden (copy editor and proofreader). This book reads and looks this good as a direct result of their involvement. Thank you for making me look good.

Finally, a big thank you and hug to my family— Carrie, Grace and Sophie—for tolerating yet another writing project. Without their support and patience none of this would ever come to fruition. I love you all.

THE MORE YOU
CAN FOCUS YOUR
TEAMS ON SATISFYING
CUSTOMER
NEEDS, EFFECTIVE
COLLABORATION,
AND CONTINUOUS
IMPROVEMENT,
IT WON'T MATTER WHICH
METHODOLOGY
THEY EMPLOY.

JEFF GOTHELF is the author of *Lean UX* and *Sense and Respond*, as well as a speaker and organizational designer. Over his 20 years in technology Jeff has worked to bring a customer-centric, evidence-based approach to product strategy, design and leadership. Jeff has worked in various roles and leadership positions, most recently as co-founder of Neo Innovation (sold to Pivotal Labs) in New York City and helped build it into one of the most recognized brands in modern product strategy, development and design. Jeff is regularly keynoting conferences, teaching workshops or working directly with client leadership teams across North America, Europe & Asia.

CPSIA information can be obtained
at www.ICGtesting.com
Printed in the USA
LVOW10s2119280217
525693LV00014B/557/P